CAREER AS A
PAINTER

PAINTING CONTRACTOR

SOME CAREERS GROW OUT OF OPPOTUNITIES you did not anticipate. People start painting houses as a part-time job while they are in school, and before they know it they have built a successful career competing for commercial painting contracts.

Painting is something that most people need. They may be able to do it themselves, but not without making a substantial investment in equipment like ladders, spray guns and compressors, not to mention brushes and rollers. Most homeowners are smart enough to know that they probably will not do the job as well as an experienced professional, and certainly not in the same amount of time. Demand for painters has never been higher, as more people renovate old homes rather than tearing them down.

There are many different kinds of painters. If painting houses does not sound right for you, how about painting enormous commercial buildings, or even infrastructure like bridges?

As with all careers, dedicated professionals rise to the top. Careful professional painters who take pride in their work will always land the best jobs with the best companies. Professionalism is the key in the painting business because it is the one thing that will make you stand out.

WHAT YOU CAN DO NOW

YOU CAN EASILY GET STARTED ON YOUR career in painting while you are still in school. When you think of a professional painter, it is probably someone who paints houses, commonly known as construction and maintenance painters. They are the most common type of painters, and demand for their services is always high because people need to maintain their homes and often change colors just to keep up with changes in fashion. They are not the only kinds of painters. Industrial painters apply protective finishes to commercial buildings, bridges, ships and other large structures. Artisanal painters work primarily in interior design and specialize in complex techniques like faux finishing, glazing, and sponge painting.

You should also consider opportunities for business ownership. Most construction and maintenance painting companies are small businesses founded by entrepreneurs who started out as painters. If you think you want to be your own boss, painting is a good way to do it.

Get some experience by painting your own house. Start with a single room. Read up on proper methods, buy the basic equipment and materials, and give it a try. Offer to start with a room that can stand up to some experimentation, like the basement or garage. Nobody will hold it against you if your first results are not perfect. As you get better, move into other rooms. Learn to use latex, the most common paint for walls, and oil-based paints commonly used for moldings and other trim pieces. They have very different characteristics and lead to equally different results.

Get a part-time or summer job with a painting contractor and put your skills to the test. You may not do much actual painting at first, but you will learn important skills like how to cover a room with drop cloths to prevent paint from getting on furniture, how to clean and maintain brushes, and how to mask trim to keep wall colors from splattering onto trim and vice-versa. These are all skills you will need to know if you want to succeed as a painter.

HISTORY OF THE CAREER

PAINTING HAS BEEN AROUND FOR thousands of years. Paint serves two basic functions: it protects the surface it is applied to and it adds color that can be decorative, practical, or both. The ability to apply color was important a long time before anybody noticed that paint protects surfaces like wood and metal. That is because there was no need to paint the walls inside a cave. Stone is already pretty sturdy, and is among the few materials paint cannot do much to protect. Cave dwellers certainly did figure out how to paint. They left thousands of cave paintings all over the world, some of which are amazingly vibrant and complex. Some paintings depict humans in family settings, while others show animals and even complete hunting scenes. Were these paintings intended to tell a story? To track the movement of herds? To make the caves more pleasant places to live? We will never know.

The earliest paints were made from whatever was available. Plants, animals and minerals provided the first pigments. Some pigments, like ochre, which is made from iron oxide, have resisted the ravages of time fairly well, leaving both traces and complete pictures behind. Other durable pigments include charcoal, hematite and manganese oxide, all of which have been identified in cave paintings and archaeological sites. In addition to pigment, paints need a base to spread the pigment and to harden into a solid surface. Egg yolks were used extensively for this purpose. Oil and water were also widely used and are the most common bases in use today.

Through the 1600s, only the very wealthy could afford to paint their homes. Making paint was a laborious manual process that required grinding pigment with a mortar and pestle, and carefully mixing it with oil and white-lead powder, commonly used to help paint solidify into a durable surface. Lead was used in paint until well into the 20th century, but was outlawed in most countries when it was discovered that lead-based paint is highly toxic. Most jurisdictions now require that existing lead-based paint be either removed or completely covered with modern paint whenever a renovation is undertaken.

By the 1700s, various inventors had devised machines to make mixing paint much easier and, therefore, less expensive. As soon as regular people could afford to paint their houses, they began to do so. Paint added a dash of color and style and, more importantly, helped wood to withstand moisture and rot. If you spent your entire life living in a wooden shack with no color, constantly replacing rotting boards, you would jump at the chance to add some colorful paint! Painting also allowed regular people to enjoy the fashions promoted by the wealthy. Their houses may have been much smaller than the mansions but at least they could be the same colors.

Paint took a great leap forward in 1866 in the United States when Henry A. Sherwin became a partner in Truman, Dunham and Company, a business that sold paint ingredients. By 1870, Sherwin had created his own business with a supplier named Edward Williams, and the Sherwin-Williams company was formed. The company revolutionized the paint business in 1878 by starting to sell high-quality, ready-to-use-paint in resealable tin cans. At the time, most homeowners and professional painters bought separate ingredients and mixed paint themselves. A few companies had offered ready-mixed paint, but quality was poor and cans could not be resealed for later use. It seems obvious today, but these simple innovations changed the paint industry forever.

Paint has been in common use ever since. Its protective qualities cannot be beat and its decorative uses have made the world a much prettier and more interesting place. Essentially all homes and other buildings are painted today. Even log cabins and other purposely rustic buildings are coated with clear finishes like polyurethane to protect their wood and help them to last longer. Modern infrastructure like bridges is always painted to keep metal from rusting, and modern steel ships and shipyard infrastructure would not even be possible without the protective qualities of modern paints. Auto painting has become a multibillion-dollar industry.

Today's paints are the result of amazing technology. Pigment combinations can create thousands of colors, and most colors can be easily mixed by a computerized mixing station at a hardware store. Paints are supplied in dozens of sheens, from flat to glossy, and every nuance in between. Paints can be glittery and can even change colors in response to changes in temperature. Paints have even been formulated to protect spacecraft from the ravages of space travel.

Painting is an important skill. Whether they are construction and maintenance painters doing relatively simple jobs on houses and small commercial buildings, or specialized industrial painters applying the latest coatings to complex structures, today's painting professionals are an important part of the modern world.

WHERE YOU WILL WORK

PAINTERS ARE IN DEMAND EVERYWHERE. Painting is a universal function for a modern society, like carpentry, electrical work and plumbing. Paints not only liven up our world with color and style, they also protect the materials they are applied to, enabling them to last longer and provide better service.

Construction and maintenance painters can find work literally anywhere. Houses, apartment buildings, and small commercial buildings need constant maintenance. New owners often like to change colors just to make their new property feel more like their own. Fashion also inspires many people to make changes to their paint schemes, as colors go in and out of style over time. Construction and maintenance painters with businesses in big cities will probably spend most of their time working in apartment buildings, while those who opt for suburban or rural communities are more likely to work mostly on single-family houses.

Industrial painters can also find work just about anywhere, but the number of opportunities is likely to be smaller than for construction and maintenance painters. Some industrial painters complete apprenticeships or earn special credentials in order to do their jobs. Industrial painting often involves exposure to large quantities of toxic chemicals and to dangerous situations like working in high places or in strange positions.

Artisanal painters tend to be in greater demand in major metropolitan areas. Artisanal painters often work alongside interior designers to help bring their artistic visions to life. Artisanal painting has much in common with construction and maintenance painting but involves more complex techniques and greater skill.

YOUR WORK DUTIES

Construction and Maintenance Painters

Construction and maintenance painters are the most numerous professional painters by far. Employed by thousands of contracting companies, small and large, in every city and town in the country, construction and maintenance painters are the skilled professionals most people think of when they hear the word "painter."

Houses, apartment buildings, and small and large commercial buildings all need skilled construction and maintenance painters to keep them in good shape, from putting on the first coats of primer and paint during construction, to freshening up the walls or changing the colors every few years for the life of the structure. Constant demand for construction and maintenance painters is driven by the need to maintain existing structures and paint new ones, and also by the dictates of fashion. Some colors and color combinations just do not age as well as others and need to be changed when a building is put on the market or when its current owner just does not like the look of it anymore.

Construction and maintenance painters usually work full time, but some prefer to work as freelancers. Some painters are part-timers or full-timers who only work for a few months of the year, such as college students working as painters over their summer vacation. Many are self-employed but never really build their own contracting companies with multiple employees and capital equipment, preferring to compete for small jobs they can do by themselves or with a partner.

Construction and maintenance painters do more than just paint. They also prepare surfaces for painting, which is a big job. For existing structures this usually means patching nail holes and other imperfections in drywall with spackling paste and drywall tape. This is a sometimes-laborious process that requires great skill. The surface has to be perfectly smooth before paint is applied or imperfections will be painfully obvious. Painters also have to prepare rooms for painting by laying drop cloths and taping off moldings, windows, outlets, thermostats and anything else that should not be painted.

Various methods of painting are used, depending upon what is being painted. Many new construction properties are sprayed, which is the fastest and least expensive way to cover a large area with one paint color. Rollers and brushes are commonly used in existing structures, with rollers used for walls and brushes used for molding and detail work. Many jobs require latex paint for the walls, which cleans up with water and is easy to use and oil-based paint for molding, which does a better job of protecting wood but has to be cleaned up with mineral spirits and is trickier to use than latex. Painters have to work as quickly and as accurately as possible.

All painters need to be mindful of customer service, even if they are not directly responsible for negotiating with the customer. Painting can be a very disruptive process, requiring people living in a house, for example, to move out of the rooms to be painted, hope the painters will take good care of the possessions they move around to make room for painting, breathe paint fumes, and generally deal with people coming and going from their house for a few days. This can be a very trying process, and all painters need to be polite and courteous.

Construction and Maintenance Managers

Most construction and maintenance painting companies put one or more experienced painters in charge of small teams. There is no formal title for these leaders, who are often known as managers or team leads. Managers typically report directly to the owner of the company and represent the owner to the client.

For example, the owner of a contracting company may meet with a client initially in order to take a look at the job to be done and to provide an estimate. After the proposal has been won, the owner can turn the job over to a manager who will make the arrangements to take a team of painters to the site. The manager is typically responsible for assembling the team, keeping track of hours, bringing the proper supplies and equipment, and buying paint. The manager also represents the owner in day-to-day dealings with the client.

Most managers start out as painters and are selected for managerial positions because they have mastered the skill and have shown that they will be good representatives of the company.

Construction and Maintenance Contractors

Construction and maintenance painting contractors are the entrepreneurs who start and run the companies that make the business go. The solid majority of painting contracting companies in the United States are relatively small businesses owned by a single owner or small partnership. Most of these small contractors work almost exclusively in relatively small areas, like a single metropolitan market, and employ fewer than 10 employees, although they may hire freelancers or other temporary employees for special projects or busy seasons.

Painting is a relatively easy business to get into. Unlike plumbing or electrical contracting, painting usually does not require expensive and time-consuming licenses and certifications. Painters do not have to wait for building inspectors to check out their work before they can proceed to the next step. Painting does not require a large capital investment in power tools or other specialized equipment. An aspiring painting entrepreneur can get started with some buckets, brushes and rollers, tarps, a stepladder or two and a van or pickup truck.

The hard parts are learning the business and building a reputation. There is much to learn about how to prepare an estimate, deal with customers, manage employees, stay within OSHA guidelines, and keep customers happy. Getting your name into the marketplace and keeping it there is another difficult step. Most small painting contractors work up relationships with property developers and real estate management companies in order to make sure they are at the top of the list when somebody needs painters in a hurry.

Like all construction professions, painting contractors need to work constantly. Say you build up a painting business with eight full-time employees. Those employees have bills to pay, just like anybody else. If you expect them to stay with your company, you need to keep them working all the time. If they are idle for a week or two you are stuck in a real jam. If they really are full-time employees, as opposed to contractors, they will want to be paid, just like other employees. If you do not have any contracts, you will not have any money to pay them. If you lay people off every time work slows down, word will get around and nobody will want to work for you again. If you pay your people even when they are not working you risk going broke.

Painting may be an excellent avenue to starting your own business but keep in mind that entrepreneurship is not for everybody. When you own the company, the buck stops with you. There will not be anybody around to take responsibility for your failings, nor will money magically appear to pay your bills. Be sure to spend a few years working for somebody else first. Learn everything you can before you take the leap.

Artisanal Painters

Artisanal painting is a specialty of construction and maintenance painters. Often called upon to work on the same jobs as construction and maintenance painters, artisanal painters apply finishing touches that most construction and maintenance painters do not know how to do. They may be self-employed freelancers, work for interior design firms, or work for construction and maintenance painting contractors.

Artisanal painters apply special finishes like glazes and crackle finishes, and handle very detailed painting like the intricacies of multicolored Victorian houses and special moldings. They may paint entire rooms, as in the case of very high-end properties with extensive detail, or they may just take on the finishing touches after the regular painters have done their work. Artisanal painters also play an important role in historic preservation, where much of the work has to be extremely precise and adhere to strict historical standards.

Most artisanal painters are highly skilled artists who have studied art. They may also paint paintings, make jewelry or ceramics, or dabble in any number of other artistic specialties. Some artisanal painters consider themselves to be interior designers, first and foremost, and learned a few specialized painting techniques in order to bring their artistic visions to life for their clients. Artisanal painters can definitely charge more for their services than regular construction and maintenance painters can, but the truth is that most houses and commercial buildings never need their services.

Industrial Painters

Industrial painters work in many different industries. A few of the industries that employ large numbers of industrial painters are public infrastructure and shipbuilding.

The painters who apply paint and other coatings to enormous suspension bridges, for example, require very specialized skills. Painters working on bridges have to learn how to work at great heights, which requires extensive training. They also use different kinds of equipment and different kinds of paint.

Painters are critical to the shipbuilding business. There is no bare metal anywhere on a ship. Ships spend all their lives in the water and any untreated metal would eventually rust. Painters working in shipbuilding learn their trade through the shipyards they work for. They have to adhere to special safety standards and become accustomed to doing things like working in oddly shaped spaces and spraying paint from small platforms dangling over the side of ships. The American shipbuilding industry is thoroughly dominated by orders for the Navy and Coast Guard. If you pursue this career you can be sure that you will have to deal with many federal regulations and that you will spray a lot of gray and white paint.

STORIES OF WORKING PAINTERS

I Am a Part-Time Construction and Maintenance Painter

"I'm a college student. Most summer jobs don't pay very well and I need to make enough money over the summer to make it possible for me to get through the school year without working. There are enough distractions during the school year. I'd rather be able to concentrate on my studies than worry about scheduling a job around my classes and homework.

Painting is fun, and it pays pretty well. I also like the fact that it gets me out and about. I don't go to the same office every day and many of our jobs are outdoors, so I look forward to those. I could be working in a store or an office, but who would want to do that during their summer vacation?

A typical day begins bright and early. The other painters and I either report to the office and ride to the job site together in a van or we drive ourselves to the site separately. It depends upon the location of the job, availability of parking, and whether or not we're all working the same hours. After we arrive, we set up the site by moving furniture, laying drop cloths and preparing surfaces to be painted. This can include scraping, spackling and priming, depending upon the surface to be painted. Then we paint.

We use brushes, rollers and sprayers, depending upon the job. Sprayers make the work go fast, but they are fussy to use and have almost no margin for error. Rollers are the preferred tool for painting walls in houses and small commercial buildings. Brushes are used mainly for trim work. There definitely is skill required. Nobody will mistake me for Picasso, but that's not the point. I can do a good job and do it in a reasonable amount of time, which are the only things that really matter.

I am majoring in business administration. I've always wanted to work for myself but I've never been sure about what I want to do. Painting may be a good way to get into owning my own business. The capital costs are pretty low, and I'm developing the technical skills with this job. I'm also making lots of connections within the business, and those should serve me well too. I still have to do an internship before I graduate. I've begun looking into internships in construction contracting so that I can learn about the contracting process. That's the ticket for me."

I Am a Construction and Maintenance Team Lead

"I put together teams of painters for specific jobs. It's up to me to look at our roster of painters and determine who has the skills necessary for each job and then schedule the team accordingly.

I've been a professional construction and maintenance painter for many years. The owner of my company selected me to be a team lead a few years ago because I showed that I could take charge, provide excellent customer service and get the job done on time and within budget. Not everybody volunteers for leadership jobs. In fact, I am still amazed by how many people would rather stay in the background than step up and take charge.

My boss does the estimates and signs all of the contracts. Then I take over. I assess the job and figure out how many painters we'll need. Then I go through my database of full-time employees and various part-timers and freelancers in order to determine who has the skills and availability to get the job done. Part of this process is purely professional. Not everybody knows how to use the spray equipment, for example, so if we need to do some spraying I need to make sure I choose the right people. The other part, however, is personal. I know some perfectly good painters who are nervous about heights and don't like to work on high scaffolds or tall buildings. That may limit their usefulness to me, but as long as there are plenty of simple jobs that don't require scaffolding everybody's happy. It takes time to get to know your people well enough to make those calls.

On the job site I am the boss's representative. I handle all dealings with the client, from discussing how to go about detail work to asking where to park or if the client is okay with my using their microwave to heat up my coffee. Constant communication with the client is essential to success.

I like my job because I enjoy the feeling of a job well done. Most jobs, it seems to me, just kind of go on and on. It's hard to step back from your desk at the end of the day and admire your work because the work never really stops. When my teams finish a job we can stand back and admire our work. I never get tired of that."

I Own a Small Construction and Maintenance Painting Company

"I like being the boss. I like to know that the success or failure of a project is ultimately up to me. There's no middleman here. I get it right or I don't. The same can be said for all entrepreneurs. We wouldn't want it any other way.

I wasn't always in the painting business. I started out working all manner of odd jobs after high school, from carpentry to auto mechanics to restaurants to flooring. Along the way I discovered that I like the residential construction business. I'm good at customer service. I want to manage projects and lead people, and I'm not keen on too many rules and regulations. Plumbers and electricians, for example, have to complete lengthy apprenticeships and maintain licenses in order to do their jobs. I'd rather not have to jump through all those hoops. I just want to run a successful business.

I started my career by getting a job as a painter with a small contracting company that did mostly residential and small commercial projects. I learned the technical skills pretty quickly but I gave myself a few years to learn the business side. I became a team lead pretty quickly, and paid close attention to everything the boss did.

When I set out on my own I started with a couple thousand dollars' worth of supplies and an old van. That was it. I hired part-timers as I needed them, until I had enough work to start hiring people on a full-time basis. The hardest part of this job, by far, is keeping the business coming. I have to have my name out there all the time. I have a website and belong to all of the referral services, like Home Advisor and Angie's List. I put a sign in the front yard when I'm working on a job and make sure to leave behind a stack of business cards when I finish a job so that the client can hand them out to their friends and neighbors.

I am able to maintain a full-time staff of 10 painters plus an office assistant. If I need more painters I can always call upon a network of local freelancers and part-timers who are happy to step up for a few days of work if the price is right. I am committed to paying my full-timers whether we are working or not, so it's in my best interest to make sure we get steady work. I try to get a little extra if possible. My painters are happiest when they're earning a little overtime.

I would recommend this career to anybody who wants to be their own boss and who has a real appreciation for the business. It's not just about painting, but about the wider world of residential and small commercial construction. I get about half of my jobs from general contractors with whom I have developed a good working relationship over the years. I even get referrals from plumbers, carpenters and electricians because we bump into each other all the time. This is a great business and I'm happy to be a part of it."

I Am an Industrial Painter at a Shipyard

"I didn't know that I was going to get into painting, I just knew that I wanted a job at the shipyard. The shipyard employs just about everybody I know. I filled out an application after high school, just to see what they might have for me.

They needed painters, so I took the job and started a four-year, in-house apprenticeship under the supervision of veteran painters who knew everything about the job. I took classes in industrial painting technology, coating technology and safety, safety, safety. Shipyards are dangerous places, especially when you're working on a ship that's still being built.

Like essentially all American shipyards, my shipyard depends upon the Navy and Coast Guard for most of its work. That means we use a lot of gray and white paint, mostly gray. Naval vessels are painted gray because it provides the best camouflage against the water and the sky, and because it's a simple color that can be touched up and repaired easily.

Ships have to be painted as they are being built. There is no other way to make sure that all the bare metal is protected. Ships spend their lives in a highly corrosive saltwater environment, so it's critically important to cover every little bit of exposed metal. Mostly we use sprayers but sometimes we have to use brushes to get into tight areas.

We use lots of other coatings in addition to paint. Aircraft carriers, for example, use a rubbery substance called non-skid on their flight decks. It helps aircraft to keep a grip on the flight deck. Without non-skid the planes would roll all over the deck, which would be dangerous. Non-skid is sturdy stuff but much of it wears off during a deployment. After six to eight months at sea the flight deck is covered by bald spots where the non-skid used to be. Those spots rust instantly and sometimes the metal has to be replaced before we can put on a new coat of non-skid for the next deployment.

I like my job because I like being a part of something much larger than myself. Applying all the paint and other coatings necessary for a ship to survive in a saltwater environment is a very important, very complicated job. I'm proud to do it."

I Am an Artisanal Painter

"Artisanal painter is a very inexact term to describe a painter with artistic skill. Some regular painters are pretty handy with a brush but don't consider themselves to be artists. I have a degree in art and also paint pictures. Working with interior designers and construction and maintenance painters is a great way to make money while still doing what I love.

I graduated from art school with a bachelor's degree and no idea about what to do to make a living. I loved painting pictures, but painting is one of those classic arts careers that are long on personal satisfaction and short on money for all but a lucky few.

I started my career as an artisanal painter almost by accident. A friend asked me to do some detail work on moldings in her house and add a glaze to a wall to make it brighten up a room. It turned out really well and led to many requests from other friends, including a few interior designers who asked about hiring me to help out with clients who wanted special paint as part of their projects. Before I knew it, I had a new career.

I rarely work alongside the regular painters. Mostly, I let them finish their part of a project before I come in and do the detail work. I also apply special finishes to entire walls, like sponge painting, crackle finishes and even custom murals. I paint a lot of animals for kids' rooms, for example.

This is freelance work. I have built up a large network of interior designers and painting contractors who call on me regularly. I also maintain my own website and work directly for clients when they need something special. I still paint pictures, of course, and do a gallery showing once in a while. I'll never give up on that, but working as an artisanal painter has been very satisfying and a great way to earn a living."

PERSONAL QUALIFICATIONS

PAINTING REQUIRES MORE SKILL THAN many people think. That is why most people hire painters rather than taking the chance on doing it themselves. Clearly, painting is not the right business if you are sloppy or messy. Nobody wants to hire a painter who doesn't pay attention to detail. You may be surprised at how fussy some clients can be, and with good reason. If you miss a spot on a stretch of crown molding because the lighting in the room cast an inconvenient shadow while you were painting, the customer will see it and demand that you come back and fix it. If you do a poor job of keeping a crisp, straight line between wall and trim colors you will also be asked to fix it. Mixing paint improperly can lead to disaster because mixing is what determines a paint's final color. There are millions of ways to make a mess! Painters must keep their work areas spotlessly clean and keep track of their tools and equipment. Dust can ruin a paint job and one misplaced bucket can turn into a tripping hazard that could knock over paint cans and create a real nightmare. Detail is very important in this business.

Painting often requires working in strange and uncomfortable positions for long periods of time. Painters also have to carry heavy equipment and paint cans around job sites and do lots of bending and stretching to get into hard-to-reach areas. This work is much easier to do if you are physically fit and put serious effort into staying that way.

All painters work for somebody else. It does not matter if you own the company or are the low person on the totem pole, you still report to your customer. That customer could be a fussy homeowner with what you think is questionable taste or a high-powered business owner or government official working on a tight deadline that you will have to meet. The only time you will not be trying to please somebody else is when you are painting your own home. Customer service does not come easily to everybody. Some people resent customers getting in the way of their business, while other people are simply shy or do not know how to deal with people. If you want to succeed in this business, you must master customer service. Your goal is always to please the customer, even if you would do things differently.

ATTRACTIVE FEATURES

ENTREPRENEURSHIP OPPORTUNITIES are just about endless, especially for construction and maintenance painters. Most construction and maintenance painters are employed by relatively small companies who compete for residential and small commercial contracts. Hours tend to be fairly flexible, pay is straightforward and may include overtime, and the distance between you and your boss will probably be minimal. Many painters who start out working for somebody else start their own businesses after only a few years. While you are working for somebody else you will have daily access to the tools of the trade, you will polish your skills and make connections within the local painting business. You will get to understand what materials and labor cost, how to work with customers about their preferences and needs, and make plenty of friends who know how to use a brush and roller. Creating a business is really just a matter of purchasing some essential supplies, a van, and legal assistance to form a corporation. Then you have to start competing for contracts, offering the best work at the lowest price. Do your best to maintain good working relationships with everybody you meet when you enter this business because you never know whose goodwill you may need in the future. Making a good impression on a general contractor when you are just starting out could pay huge dividends in the future when that general contractor offers you a subcontract on a big job. If you have ever wanted to be your own boss, construction and maintenance painting offers an excellent way to go about it.

Construction and maintenance painting is not going to make you Rembrandt, but it is still a creative process. You may be amazed at how little most people know about color combinations and how to go about painting a house. Did you know there are dozens of shades of white commonly used in painting houses? Or that choosing many different colors or types of paint can dramatically increase labor costs? Most of your clients will not know these things. It will be your job to help them achieve the look they want at the price they are willing to pay. This can be more creative than you may imagine. Getting a color combination right is not easy. Finishing a job, standing back and admiring your work and knowing that you nailed it is a very satisfying experience. The process of painting can be fun, too. This is not a desk job. You can look forward to being out and about all day.

Painting is also a relatively secure business. More and more people nowadays see the value in preserving existing homes and commercial buildings. Preserving existing buildings is a form of recycling that has significant benefits for all concerned, from conserving energy and raw materials to brightening neighborhoods and preserving architectural heritage. Demand for painters does dip a bit when the economy dives, but not as much as many other industries. Preserving an existing structure is almost always less expensive than building a new one. That is appealing in good times and bad.

UNATTRACTIVE ASPECTS

TURNOVER IN THIS BUSINESS IS HIGH because the barriers to entry are so low. Working for a small construction and maintenance painting company is a little like working for a restaurant. Many of the people working in both businesses have their sights set on something else and tend to move along as soon as they are able. Painting is a popular choice among college students for part-time and summer employment, for example. The business also attracts large numbers of newly arrived immigrants looking to make some money while they work on their English and earn a credential to do something else. At the entry level, painting requires very little skill, making it a natural choice for people without skills or who just want to pay the rent while they pursue something else. This does create headaches for small-business owners who have to constantly hire and train new employees as their staff turns over.

Competition is fierce between construction and maintenance painting companies. This is due to many factors. Because entry-level painters require so few skills they can be paid relatively little. Some companies resort to paying their painters in cash, especially if they hire many undocumented immigrants, because it gives them some wiggle room when competing for contracts (paying in cash and hiring undocumented immigrants are both illegal but, unfortunately, quite common). Sometimes most of the painting companies in an area get booked up for weeks at a time because a large developer needs to have an entire subdivision painted all at once. That leaves the remaining companies free to charge more for other jobs because they are the only game in town while everybody else is booked.

Painting is not exactly a dangerous activity in the way that, say, firefighting is. Painters rarely suffer serious injuries on the job. If you get into this business, you will suffer minor injuries. You will spend your days hauling ladders, compressors and other pieces of heavy equipment to and from trucks, setting up scaffolding, and bending and stretching into impossible positions. You will also breathe paint fumes all day, every day. Get used to wearing a respirator. Painters have to live with cuts, scrapes, pulled muscles and minor body aches pretty much all the time. These are relatively minor problems, however. The only exception to the minor-injury rule is the danger posed by working at great heights. Once in a while a painter falls off a roof or a scaffold collapses, resulting in serious injury.

EDUCATION AND TRAINING

JUST BECAUSE THERE ARE NO EDUCATION requirements to get into the painting business does not mean that you should not continue to pursue formal education after you graduate from high school. You are not going into painting just as a summer job, but as a career.

The first thing you must do is take shop classes in high school. Woodworking is especially important for aspiring construction and maintenance painters, while metalworking will be more useful for those aiming for a career in industrial painting. Shop classes may not put much emphasis on painting, but they will teach you the basics. They will also teach you many associated skills, like basic carpentry and metalworking. Shop classes will introduce you to the world of what are commonly known as the trades, that cluster of professions that includes everything from carpentry and plumbing, to painting and auto mechanics. You will develop good safety habits that will serve you well later in your career. Working in a shop or on a construction site can be dangerous. The more time you spend in that environment, the better.

Many painters and careerists pursuing related trades earn associate degrees in business administration or in a more-specific major like construction trades. Such degrees are an excellent way to learn business math, how to estimate the cost of a project, how to manage people, and how to run your business in accordance with local laws and regulations. Nobody sorts out Occupational Safety and Health Administration – OSHA – regulations without some help. Most community college schedules are created with working adults in mind, making it possible to earn a degree by taking a class or two at a time while you are working.

There are also apprenticeship programs for painters. The International Union of Painters and Allied Trades, for example, offers specialized apprenticeships in commercial painting and wallcovering, drywall finishing and installation, bridge and industrial painting, and glazing and architectural metal installation. All apprenticeships require three to four years of on-the-job training and a minimum of 144 hours of classroom instruction. Specialized industries offer their own apprenticeship programs. Major US shipyards like Huntington Ingalls Industries, for example, offer their own apprenticeship programs for careerists interested in specialized industrial painting careers.

If your goal is to start your own business someday, you should seriously consider pursuing a bachelor's degree in business administration. A bachelor's degree will not only put you at the head of the pack for leadership positions within somebody else's company,   it will also give you more leverage later in your career when the time comes to apply for a commercial loan to start or expand your own business. A bachelor's degree in business administration will give you a well-rounded foundation in accounting, marketing, finance, advertising, management and leadership. You will need all of these skills as you pursue your career as a painting contractor and business owner.

One thing you should be sure to take advantage of during your college years is an internship. An internship is a full-time job related to your major that takes the place of classes for a summer or semester. Most internships are paid, and many come with special opportunities for interns not normally offered to other employees, like meetings with senior staff or special trips to project sites. An internship with a construction company, for example, will give you insight into the business you could not hope to get from inside a classroom. You will work alongside experienced professionals and get to look over their shoulder when they make decisions. Never again will you have the opportunity to try on a career for a few months and then change course if you decide it is not right for you. Best of all, many recent college graduates get their first full-time jobs after college with the companies where they worked as interns. Go to your school's internship advisor and see what is available.

EARNINGS

EARNINGS CAN VARY QUITE A BIT IN the painting business, from part-timers working on summer jobs to painting contractors winning major contracts. Most painters fall somewhere into the middle.

First, the low end. Entry-level painters and careerists just passing through on their way to something else can expect to earn about $25,000 to $30,000 per year if they work full time. Very expansive markets like San Francisco or New York City may pay a little more. Serious careerists who work full time for an established construction and maintenance painting company can expect to earn more like $35,000 to $45,000 per year, and possibly as much as $65,000 or even $70,000 per year if they step into managerial positions and lead teams on projects. Painters working for industrial painting companies or in specialized industries like shipbuilding tend to earn more than construction and maintenance painters but generally have to maintain specialized certifications.

Big earnings can be achieved when owning and operating your own painting business. Painting is a classic build-it-yourself business. You will start by working for somebody else, make a few connections, invest in some tools, land a job or two, hire an extra set of hands and within a few years own your own full-fledged business. In fact, about 40 percent of construction and maintenance painters are self-employed. Some of these entrepreneurs are basically freelancers hiring themselves out to existing companies while others are the founders and owners of small businesses competing for multiple contracts. This degree of flexibility is a major bonus to the painting business.

Earnings of business owners can vary depending on the size of the company and how successful you become. The range can easily reach $100,000 or more, when you become well established.

Painting is a career you can make your own. Full time or part time, as an employee, freelancer, or business owner. Few careers offer so many options. Do not be put off by the low entry-level pay. The amount of money you can make is limited only by your own determination to make it.

OPPORTUNITIES

THERE ARE MANY WAYS TO MOVE UP in a painting career, most of which entail stepping up to greater responsibility. The painting business is filled with adequate painters who get the job done well enough to keep customers reasonably happy. These painters may or may not stay in the career for the long term. The best painters rise to the top very quickly. If you can do an excellent job, and do it ahead of schedule and on budget, you will quickly be put in charge of small teams and get a raise. Separating yourself from the pack of ordinary workers will help to propel your career forward. This also goes for basic professionalism like showing up on time, cleaning up after yourself, and treating customers with the attention and respect they deserve. Take your job seriously and you will be rewarded.

The best painters will be asked to move into managerial positions. The first step is usually as a team supervisor leading small groups of painters working on specific contracts. Team leads make sure everybody shows up on time, that everybody has the paints and equipment that they need, and that work wraps up on time. Team leads are typically also responsible for keeping track of hours worked and relaying that information to the office. They may also be the first line of responsibility for training new employees and counseling employees who are having troubles on the job. Successful team leaders may move up to become managers responsible for planning and scheduling jobs, and even writing up estimates and competing for contracts. As in other businesses, managers are paid more in exchange for taking on greater responsibilities.

Starting your own business is the ultimate way to the top of the painting profession. A construction and maintenance painting company can take many forms. Some entrepreneurs are content to build a small company that can handle one or two contracts at a time. Advantages to small companies include small staffs with relatively few labor issues, the ability to wear multiple hats and keep things straight, and near total control over the business. If your goal is a steady income and a predictable schedule, a small painting business could be your ticket to a comfortable life. There is no doubt that larger businesses can ultimately make more money, especially if they can juggle many contracts simultaneously or invest in specialized equipment like industrial paint sprayers and complex scaffolding systems enabling them to tackle very large jobs. Big businesses are complex and even the most dedicated entrepreneur cannot control all aspects of a large enterprise engaged in multiple contracts. You will have to decide which way you want to go.

GETTING STARTED

GETTING STARTED ON YOUR CAREER IN painting is a pretty straightforward process. First, get in touch with every connection you have ever made in the painting business and offer your services. When the time comes to start your career on a full-time basis, you may already be working part time as a painter or have some recent connections in the business from summer jobs. Seek out your connections in the business and spin up a full-time job. Entry-level painting jobs are typically secured by word-of-mouth and personal recommendations, but having a ready-to-go résumé will help. There are many books and software applications that can help you to put together a polished, professional résumé that will showcase your experience and credentials in the best possible light. Keep paper copies with you at all times and save the digital file on your laptop so you can attach it to an email in a split second. Include all of your experience, and be sure to point out any leadership or managerial responsibilities.

As soon as you have locked down a full-time job, make yourself stand out from the crowd by volunteering for the hard jobs. This can mean many things: big jobs, inconvenient jobs, dangerous jobs, jobs with lots of responsibility. Step up! Bosses love employees they can count on to tackle the difficult jobs nobody else wants. Demonstrating that you are the go-to person for hard jobs will move you up the ladder faster than anything else. The sooner you get serious managerial experience, the sooner you can think about starting your own business, if that is what you want to do.

Not every job has to be a dream job. Your first full-time job needs to give you enough experience to get you to the next level. Learn what you can and move on as soon as possible. Painting may very well be your life-long career, but that does not mean you have to set your sights on a life-long job. Keep an open mind and learn everything you can. Good luck!

ASSOCIATIONS, PERIODICALS, WEBSITES

■ **American Coatings Association**
www.paint.org

■ **Angie's List**
www.angieslist.com

■ **Benjamin Moor**
www.benjaminmoore.com

■ **Building Trades Education Service**
www.buildingtradeeducation.com

■ **Coatings World**
www.coatingsworld.com

■ **Finishing Contractors Association**
www.finishingcontractors.org

■ **General Dynamic**
www.gd.com

■ **Greenworks Painting**
www.greenworkspainting.com

■ **Home Advisor**
www.homeadvisor.com

■ **Home Builders Institute**
www.hbi.org

■ **Home Depot**
www.homedepot.com

■ **Huntington Ingalls Industries**
www.huntingtoningalls.com

■ **Industrial Painter**
www.industrialpainter.com

■ **International Union of Painters and Allied Trades**
www.iupat.org

■ **Lowe's**
www.lowes.com

■ **Master Painters Institute**
www.mpi.net

■ **McCormick Paints**
www.mccormickpaints.com

■ **National Center for Construction Education and Research**
www.nccer.org

■ **Painters USA**
www.paintersusainc.com

■ **Painting and Decorating Contractors of America**
www.pdca.org

■ **Paint and Decorating Retailers Association**
www.pdra.org

■ **Painters Chatroom**
www.painterschatroom.com

■ **Paint Info**
www.paintinfo.com

■ **Painting Pro Times**
www.paintingprotimes.com

■ **Paint Quality Institute**
www.paintquality.com

■ **Paint Talk**
www.painttalk.com

■ **PPG Porter Paints**
www.ppgporterpaints.com

■ **Sherwin-Williams**
www.sherwin-williams.com

■ **Signatory Painting Contractors Organization**
www.spco.org

■ **Society for Protective Coatings**
www.sspc.org

■ **Society of Decorative Painters**
www.decorativepainters.org

■ **Williams Professional Painting**
www.williamsprofessionalpainting.com

Copyright 2016 Institute For Career Research
Careers Internet Database Website www.careers-internet.org
Careers Reports on Amazon
www.amazon.com/Institute-For-Career-Research/e/B007DO4Y9E
For information please email service@careers-internet.org